Principle of Time

Randy Kubick

Preamble - In theoretical physics assumptions must be made. But never assume all assumptions are correct. When Einstein derived kinetic time dilation he assumed that any frame that is inertial can claim they are stationary. Thus, two inertial frames moving relative to each other can both claim they are stationary and it's the other frame moving. Einstein thus assumed each respective stationary frame would observe the same velocity v for the other moving frame. So let's assume both these assumptions are slightly askew and see where it takes us with our primary objective to understand why time slows down with speed.

Introduction - Just like it is for the Doppler effect, the causing mechanism of kinetic time dilation is matter moving relative to the emf while interacting with its waves (light). Knowing this will allow us to show that the rate at which time passes is directly proportional to the distance light travels in the emf between its interaction events (i-vents) with matter. An i-vent is when matter emits, absorbs, or reflects light creating an event-point coordinate in the emf.

Corroborated Assumption - Let's assume the emf surrounding earth is stationary relative to earth. While the causing mechanism is unknown, both the GPS satellites (moving) and the 1887 Michelson-Morley experiment (stationary) corroborate this assumption.

Figure 1 shows transverse and longitudinal cycling photon-clocks.

Transverse Kinetic Time Dilation Formula

$$t_{stationary} = \frac{t_{transverse}}{\sqrt{1-\frac{v^2}{c^2}}} \qquad \text{eq. 1}$$

Longitudinal Kinetic Time Dilation Formula

$$t_{stationary} = \frac{t_{longitudinal}}{1-\frac{v^2}{c^2}} \qquad \text{eq. 2}$$

Deriving Longitudinal Kinetic Time Dilation - If it takes c $1s$ to traverse a photon-clock (p-c) with side length $L = 1ct$, how can we increase $c's$ elapsed time to $2s$ without changing the p-c's dimensions or $c's$ speed? Answer: double the quantity of emf in the p-c. How do we double the quantity of emf in the p-c? Answer: give the p-c a velocity of $0.5c$ relative to its emf. Thus, a p-c moving at $0.5c$ relative to its emf will displace $2ct$ of emf while c simultaneously traverses its length $L = 1ct$ thereby doubling $c's$ elapsed time from $1s$ (stationary) to $2s$ (moving). See eq. 4. See **Figure 2**.

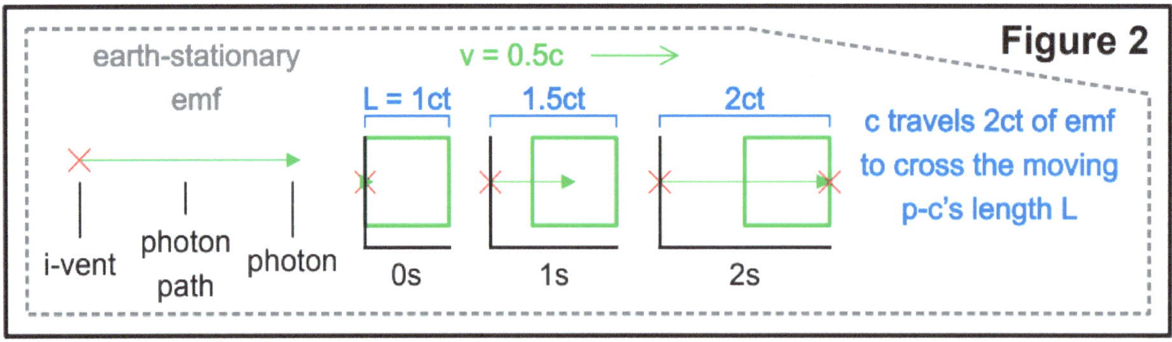

Because matter (mass/Higgs field) and the emf do not interact (minus i-vents), a p-c can move independent of and relative to its emf and its cycling photon thereby increasing and decreasing the cycling photon's field-distance between its i-vents. This, respectively, causes kinetic time dilation (eq. 4) and kinetic time contraction (eq. 6).

Eq. 3 is $c's$ net field-distance traversing the p-c's length L while it's moving at velocity v relative to the earth-stationary emf.

$$ct = L + vt \qquad \text{eq. 3}$$

Solve for time t.

$$t = \frac{L}{c-v} \qquad \text{eq. 4}$$

Eq. 5 is $c's$ net field-distance coming back across the moving p-c.

$$ct = L - vt \qquad \text{eq. 5}$$

Solve for time t.

$$t = \frac{L}{c+v} \qquad \text{eq. 6}$$

Adding the right hand side (rhs) of eq. 4 & 6 gives us $c's$ net cycle time in the moving p-c.

$$t_{net} = \frac{L}{c-v} + \frac{L}{c+v} \qquad \text{eq. 7}$$

$c's$ net cycle time when the p-c is emf-stationary.

$$t_{stationary} = \frac{2L}{c} \qquad \text{eq. 8}$$

Set the rhs of eq. 8 equal to the rhs of eq. 7 with the emf time-distance factor t_l on the left hand side (lhs).

$$t_l \frac{2L}{c} = \frac{L}{c+v} + \frac{L}{c-v} \qquad \text{eq. 9}$$

Solve for t_l to get the longitudinal kinetic time dilation formula.

$$t_l = \frac{1}{1-\frac{v^2}{c^2}} \qquad \text{eq. 10}$$

Discussion - Eq. 3 says $c's$ net field-distance to traverse the length of the moving p-c is greater than the clock's length L. This is because the p-c is moving relative to its emf and its cycling photon which increases the photon's net field-distance from L (stationary) to $L + vt$ (moving) thereby causing kinetic time dilation (eq. 4). See **Figure 3**. Note that there's four separate reference frames in **Figures 3** & **4** - the stationary, the moving, the photons, and the emf.

Reciprocal Corollary of the Constancy of Light - Gravity notwithstanding, simultaneously emitted photons must travel equal distance independent of their path.

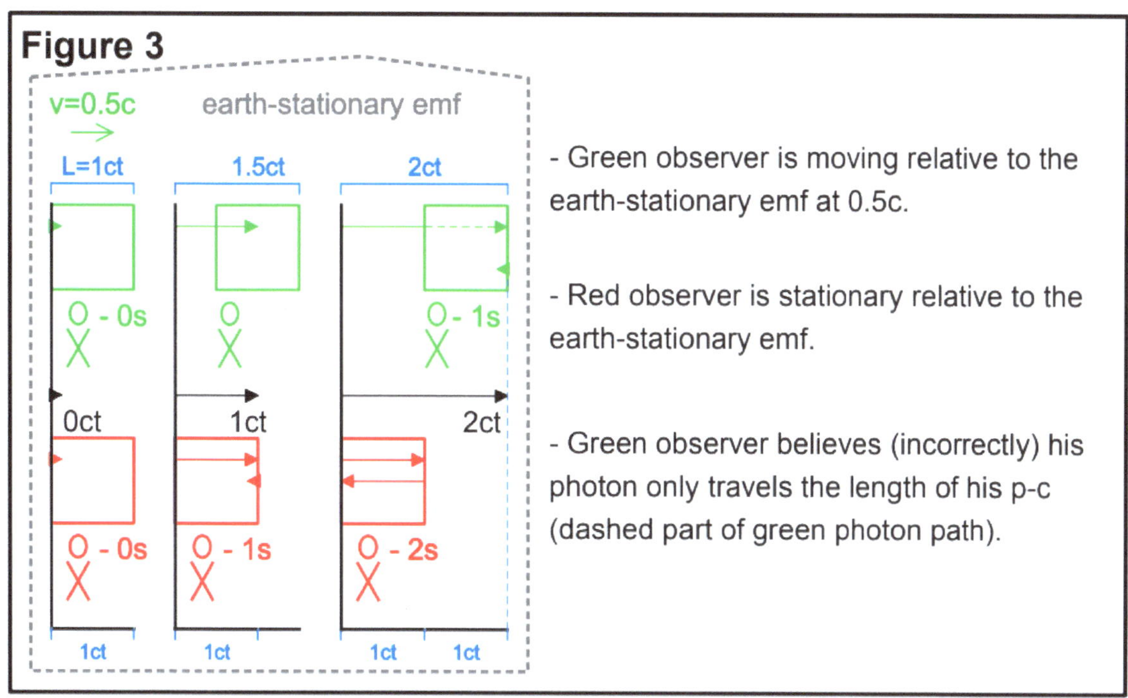

Eq. 5 says $c's$ net field-distance to return back across the moving p-c is less than the clock's length L. This is because c reversed direction at its right i-vent and is now moving back towards the left wall of the oncoming p-c. This decreases $c's$ net field-distance from L (stationary) to $L - vt$ (moving) thereby causing kinetic time contraction (eq. 6). See **Figure 4.**

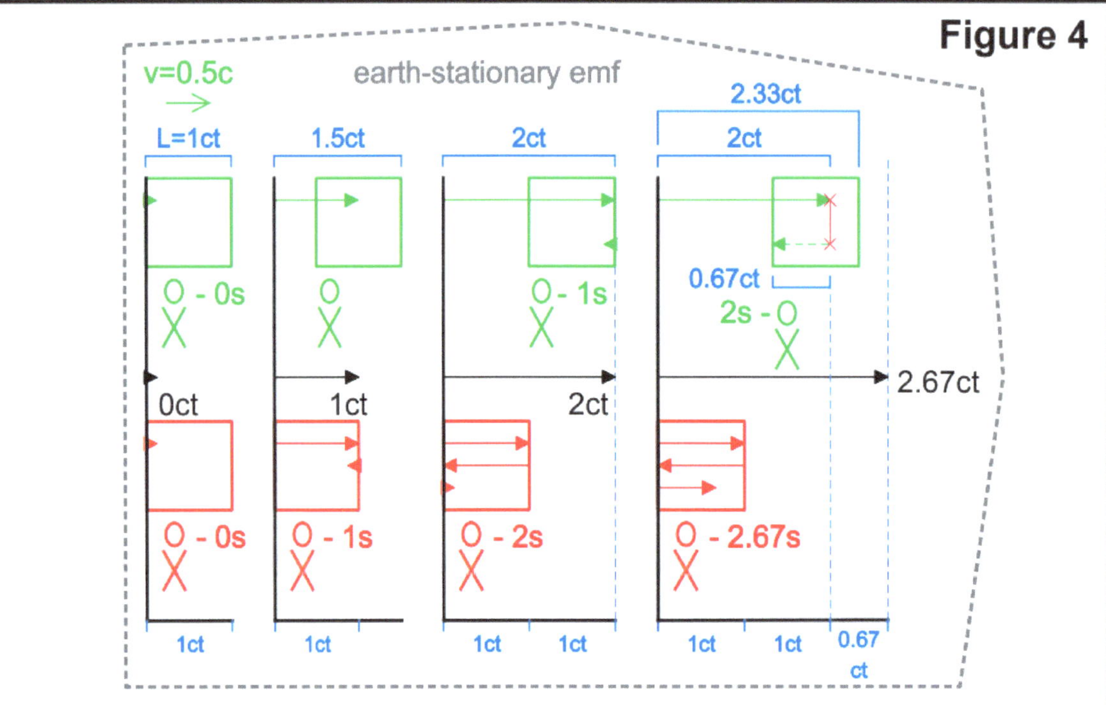

- Green observer believes (incorrectly) that his returning photon traveled the full length of his p-c (dashed part of green photon path).
- Red observer displaces 1 Kubick to complete one p-c cycle.
- Green observer at v=0.5c displaces 2.33 Kubicks to complete one p-c cycle.
- 1 Kubick = 1 cube (side length 1ct) of earth-stationary emf.

Adding eq. 3 and eq. 5 we see c's net field-distance to complete one cycle in its p-c increases from $2L$ (stationary) to $(L + vt) + (L - vt)$ (moving). This increased field-distance thereby increases the p-c's net cycle time (eq. 4 + eq. 6) in direct proportion. Thus, we see that kinetic time dilation in a longitudinal p-c is directly proportional to the photon's net electromagnetic field-distance between its respective cycling i-vents, just like it is in a transverse moving p-c.

Longitudinal - Let's show c is always c for a longitudinal cycling p-c. The net cycling field-distance of c in a moving longitudinal p-c is (eq. 3 + eq. 5) while its net elapsed cycle time is (eq. 4 + eq. 6). Since c is constant (eq. 3 + eq. 5) must equal (eq. 4 + eq. 6) so that c = c for any velocity v.

$$let\ c = 1 \rightarrow c = \frac{d}{t} \rightarrow c = \frac{eq\,3 + eq\,5}{eq\,4 + eq\,6} \rightarrow \frac{(L+vt)+(L-vt)}{\left(\frac{L}{c-v}\right)+\left(\frac{L}{c+v}\right)} \rightarrow$$

$$t = \frac{L}{c-v}, t = \frac{L}{c+v} \rightarrow \frac{\left(L+v\left(\frac{L}{c-v}\right)\right)+\left(L-v\left(\frac{L}{c+v}\right)\right)}{\left(\frac{L}{c-v}\right)+\left(\frac{L}{c+v}\right)} = 1 = c \qquad \text{eq. k2}$$

Transverse - Let's show c is always c for a transverse cycling p-c. See **Figure 5**.

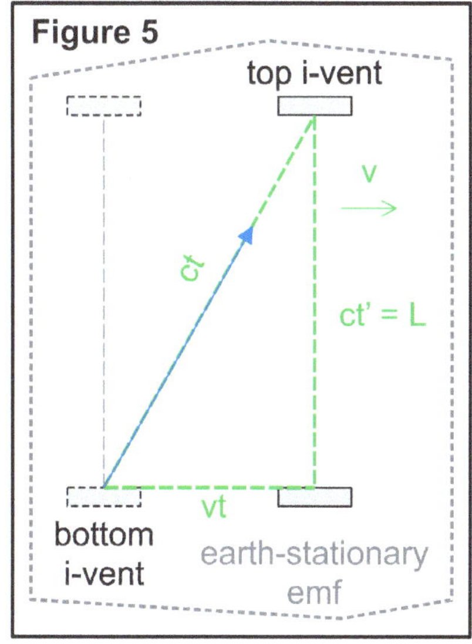

Eq. 11 is the transverse kinetic time dilation equation.

$$c^2t^2 = v^2t^2 + L^2 \qquad \text{eq. 11}$$

Solve eq. 11 for time t. This is the elapsed time for c to traverse from the bottom i-vent to the top i-vent as the p-c moves transversely at velocity v relative to the earth-stationary emf.

$$c^2t^2 = v^2t^2 + L^2 \rightarrow t^2 = \frac{L^2}{c^2-v^2} \rightarrow t = \sqrt{\frac{L^2}{c^2-v^2}} \qquad \text{eq. 12}$$

Solve eq. 11 for ct. This is $c's$ field-distance from the bottom i-vent to the top i-vent as the p-c moves transversely at velocity v relative to the earth-stationary emf.

$$c^2t^2 = v^2t^2 + L^2 \rightarrow ct = \sqrt{v^2t^2 + L^2} \qquad \text{eq. 13}$$

Eq. 13 give us $c's$ field-distance and eq. 12 is $c's$ elapsed time to traverse this field-distance. Because c is constant, eq. 13 must equal eq. 12 so that c = c for any velocity v.

$$let\ c = 1 \rightarrow c = \frac{d}{t} \rightarrow c = \frac{eq\ 13}{eq\ 12} \rightarrow \frac{\sqrt{v^2t^2+L^2}}{\sqrt{\frac{L^2}{c^2-v^2}}} \rightarrow$$

$$t = \sqrt{\frac{L^2}{c^2-v^2}} \rightarrow \frac{\sqrt{v^2\sqrt{\frac{L^2}{c^2-v^2}}^2+L^2}}{\sqrt{\frac{L^2}{c^2-v^2}}} = 1 = c \qquad \text{eq. k1}$$

Anisotropic - Because time-rate is directly proportional to the field-distance between $c's$ i-vents, time-rate is path dependent and therefore anisotropic. Thus, a moving p-c can use the difference between eq. 1 and eq. 2 to determine its speed relative to its emf. Though it appears we are violating the principle of relativity - which requires a second reference frame to determine one's speed - we're not. When we emit a photon inside a moving p-c, the photon leaves the p-c (mass/Higgs field) and emerges in the emf. Thus, the moving p-c now has the emitted photon's emf path and its respective emf-stationary i-vents as a means to determine its speed relative to its emf. Hence, because matter can move relative to its emf while c is a constant in it, a matter particle's emf is its default stationary reference frame. When eq. 1 = eq. 2 a p-c is stationary relative to its emf (1887 Michelson-Morley).

emf-stationary - the fastest decay rate speed - Gravity notwithstanding, a particle's fastest decay rate is when it's emf-stationary - ie - not moving relative to its respective emf. But once it starts moving relative to its emf, its decay rate will only track slower with increasing speed (see **Figure 3**). This means a particle's decay rate is a function of its speed relative to its emf - not another particle. Thus, the twin paradox isn't possible because the decay rate of each respective twin is a function of their speed relative to their respective emf - not each other. In other words, assuming a single earth universe and ignoring gravity, the earth-stationary twin will always have the fastest decay rate because she is stationary relative to the earth-stationary emf. Any speed above this zero-speed relative to earth will only track slower - ie - the twin moving relative to earth and its respective earth-stationary emf.

Add Speed, Add Field, Add Time - Accuracy notwithstanding, all events are clocks and all clocks are events. So if the decay rate of a particle (an event) is 100 seconds when stationary, then its decay rate for an observer moving with it at velocity v is also 100 seconds. This is because the observer's moving p-c (an event) slows at the same rate as the decaying particle (an event). But understand that the cycling photon inside the moving p-c does not slow down. It moves independent of and relative to the p-c while the p-c moves independent of and relative to it (Doppler effect). Thus, a p-c, when moving relative to its emf displaces a greater quantity of emf, and because $c's$ speed is constant in the emf, the increased quantity of emf (see **Figure 2**) causes a longer (not slower) elapsed p-c cycle time. So time doesn't actually slow down with speed - the p-c moving relative to the emf simply adds emf between its i-vents which in turn adds time.

Principle of Time - Because time equations 4, 6, & 11 all derive from distance formulas, we can say a p-c's cycle time is equal to its photon's cycle distance. So if time is the principle of change, then its rate is a function of the electromagnetic-field-distance between $c's$ i-vents. Or more fundamentally, time-rate is a function of the field quantity between a force carrier particle's i-events.

Gravity - Is the causing mechanism of gravitational time dilation different than the causing mechanism of kinetic time dilation? If the elapsed time of an event (photon-clock) is longer at the bottom of a gravity well than it is at the top, could we argue it's due to a greater quantity, or higher density of emf at the bottom of the gravity well than at the top? If we were to take 2 cube units of earth-stationary emf (side length $1ct$) and combine them into 1 cube unit, would the elapsed time of c across this 2X denser cube of emf double to $2s$? See **Figure 6**.

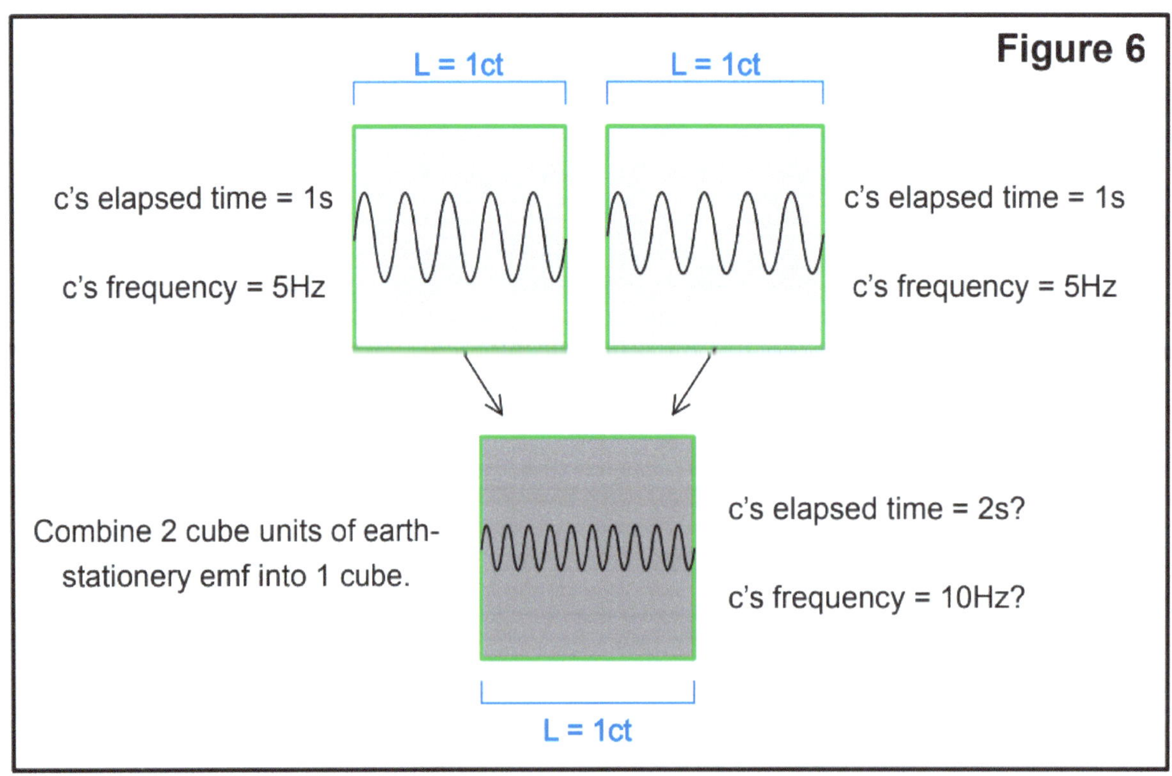

Transformation Underpin - Since eq. 1 was derived from Galilean transformations let's derive eq. 3 and eq. 5 from their respective individual transformations to underpin eq. 2.

Standard Configuration - Consider two isotropic inertial frames, S and S'. S is earth-emf-stationary while S' moves at velocity v relative to S in the positive x direction. Frame S notation is (x, y, z, t) while S' is (x', y', z', t'). At $t = t' = 0$, S and S' origins are transposed, and a photon, emitted from the transposed origins propagates in the positive x-axis direction. Because frames S and S' are isotropic, the photon has no displacement in either the y, y', z or z' axes, so photon displacement in y & y' = 0 and z & z' = 0. Thus, the y, y', z and z' axes have no bearing on our x and x' axes longitudinal transformation equations. See **Figure 7**.

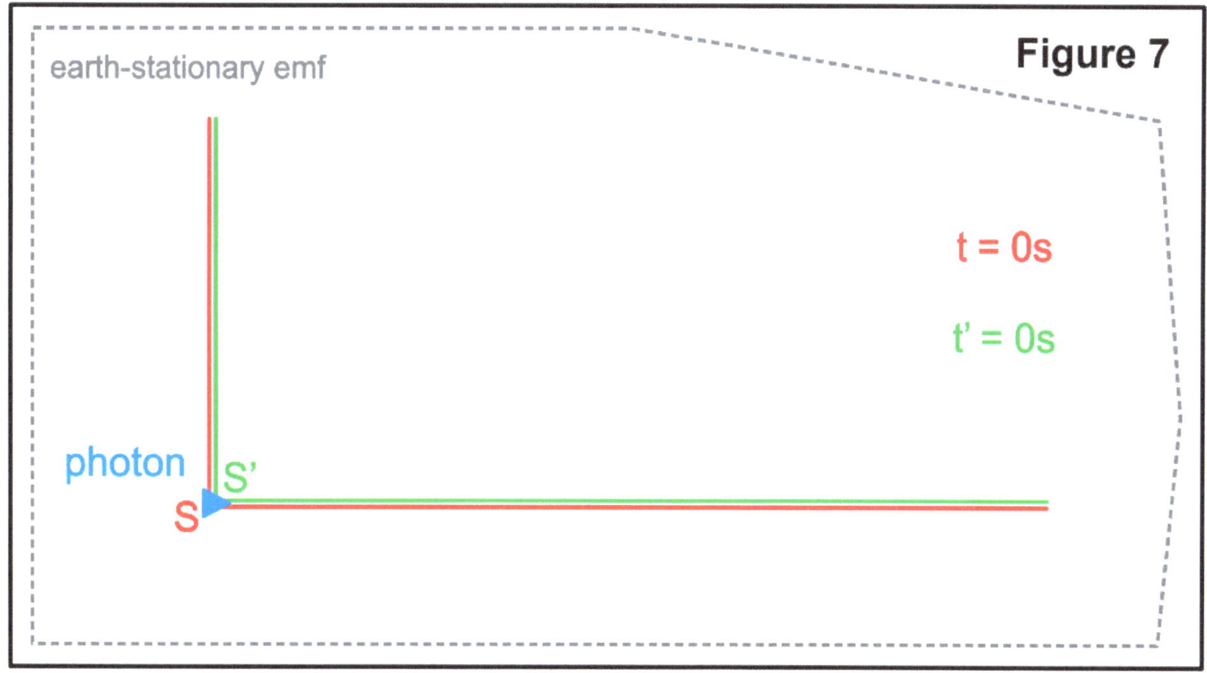

After $1s$ has elapsed in the S frame with the S' frame moving at $0.5c$ along the positive x-axis. See **Figure 8**.

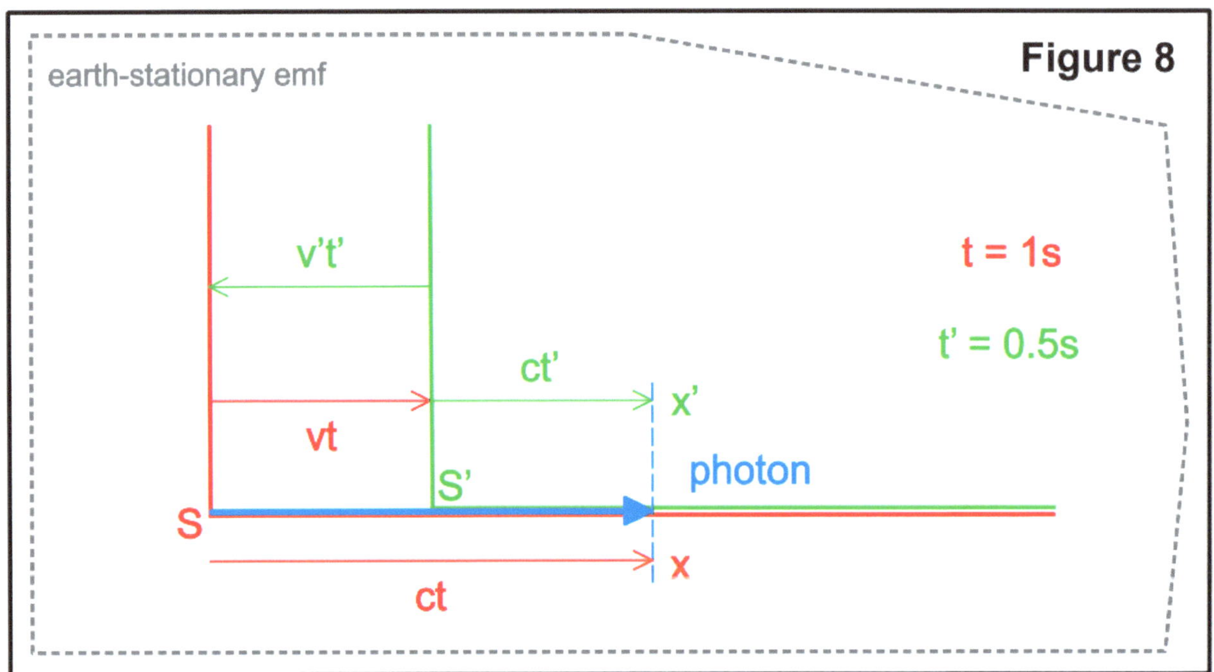

Stationary S Red says

$$S' = vt \qquad \text{eq. 14}$$

$$x = ct \qquad \text{eq. 15}$$

Moving S' Green says

$$S = v't' \qquad \text{eq. 16}$$

$$x' = ct' \qquad \text{eq. 17}$$

When eq. 16 is expressed from the S' frame, v must be v' otherwise t and t' will not divide out when multiplying out vt'. The product of vt' will also give an incorrect distance calculation for the S' Green observer thereby violating the Lorentz invariance principle.

$$vt' = \frac{m}{t}t' \quad verses \quad vt = \frac{m}{t}t \quad \& \quad v't' = \frac{m}{t'}t' \qquad \text{eq. 16}$$

Let's find the time factor difference between S and S' by setting the rhs of eq. 15 equal to the sum of the rhs of eq. 14 & 17

$$ct = ct' + vt \qquad \text{eq. 18}$$

Solve for t

$$ct - vt = ct'$$

$$t(c - v) = ct'$$

$$t = \frac{ct'}{(c-v)}$$

$$t = \frac{t'}{\left(1-\frac{v}{c}\right)} \qquad \text{eq. 19}$$

Because the events in the **Figure 8** transformation are actually the same as those in **Figure 3** (minus the p-c box), eq. 3 must equal eq. 18. Replace moving L with ct' in eq. 3 and we see it equals eq. 18.

$$ct = L + vt \rightarrow ct = ct' + vt \qquad \text{eq. 18}$$

So we've shown eq. 3 = 4 = 18 = 19 but we have only derived half of eq. 2. Let's show when a photon is emitted in the opposite direction of S' the resulting x-axis longitudinal transformation equations equal eq. 5 - the other half of eq. 2. See **Figure 9 & 10**.

Note in **Figures 9 & 10**, because we are calculating net elapsed times and net displacements and not coordinates or vectors, the negative sign used for negative x-axis coordinates and vectors must not be used as it will lead to incorrect displacement calculations.

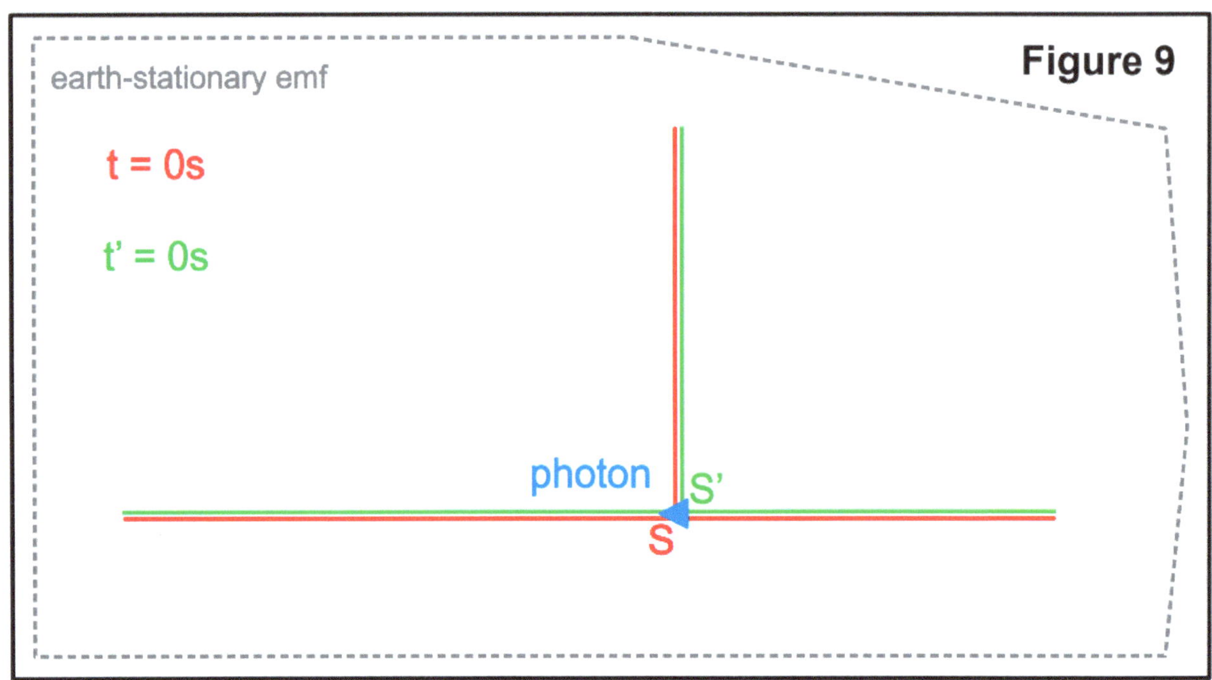

After $1s$ has elapsed in the S frame with the S' frame moving at $0.5c$ along the positive x-axis See **Figure 10.**

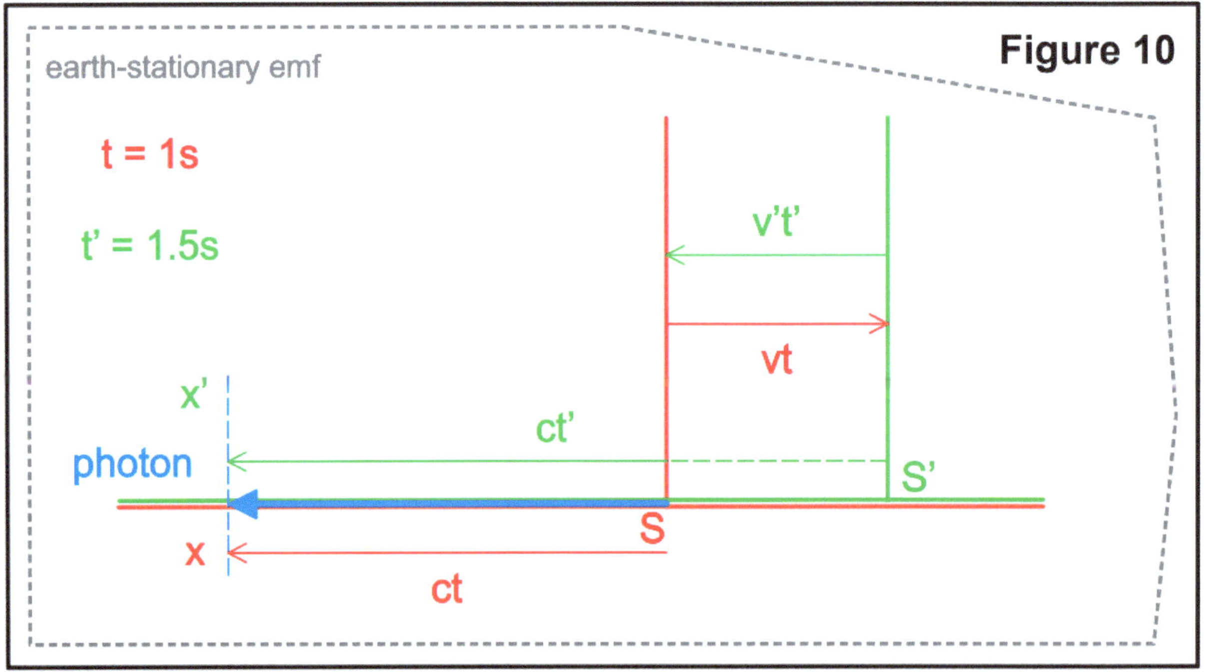

Stationary S Red says

$$x = ct \quad \text{eq. 20}$$

$$S' = vt$$ eq. 21

Moving S' Green says

$$x' = ct'$$ eq. 22

$$S = v't'$$ eq. 23

Let's find the time factor difference between S and S' by setting the rhs of eq. 22 equal to the sum of the rhs of eq. 20 & 21. (Note - because S' Green is moving inertially relative to the earth-stationary emf, he incorrectly believes the photon's emission point is at his origin while it's actually at S Red's origin).

$$ct' = ct + vt$$ eq. 24

Solve for t

$$ct + vt = ct'$$

$$t(c + v) = ct'$$

$$t = \frac{ct'}{(c+v)}$$

$$t = \frac{t'}{\left(1+\frac{v}{c}\right)}$$ eq. 25

Because the events in the **Figure 10** transformation are actually the same as those in **Figure 4** (beginning at the right i-vent), eq. 5 must equal eq. 24. Replace moving L with ct' in eq. 5 and we see it equals eq. 24.

$$ct = L - vt \rightarrow ct = ct' - vt \rightarrow ct' = ct + vt$$ eq. 24

In summation, eq. 3 = 4 = 18 = 19 and eq. 5 = 6 = 24 = 25. **Figures 3 & 4** (from right i-vent) **= 8 & 10** respectively.

The Missing Prime - So why doesn't eq. 2 equal eq. 1 if Lorentz-Einstein derived eq. 1 from the same **Figure 8** longitudinal transformation? And how did they derive a transverse kinetic time dilation equation from a longitudinal transformation? Let's look at Lorentz-Einstein derivations and see if we can derive eq. 1 from just the **Figure 8** transformation.

$$t_{stationary} = \frac{t_{longitudinal}}{1-\frac{v^2}{c^2}} \qquad \text{eq. 2}$$

$$t_{stationary} = \frac{t_{transverse}}{\sqrt{1-\frac{v^2}{c^2}}} \qquad \text{eq. 1}$$

In **Figure 8** Lorentz says the photon distance from S origin to the photon coordinate x can be calculated with Pythagorean's theorem.

$$x^2 + y^2 + z^2 = c^2t^2 \rightarrow x^2 + y^2 + z^2 - c^2t^2 = 0 \qquad \text{eq. 26}$$

Using Pythagorean's theorem again to calculate the photon distance from S' origin to the photon coordinate x'.

$$x'^2 + y'^2 + z'^2 = c^2t'^2 \rightarrow x'^2 + y'^2 + z'^2 - c^2t'^2 = 0 \qquad \text{eq. 27}$$

Because eq. 26 and eq. 27 both express the same photon coordinate we can equate them as follows.

$$x^2 + y^2 + z^2 - c^2t^2 = x'^2 + y'^2 + z'^2 - c^2t'^2 \qquad \text{eq. 28}$$

Because S and S' frames are isotropic and our transformation only occurs along the x-axis, photon displacement along the y & y' = 0 and z & z' = 0. Thus, we can simplify eq. 28 as follows.

$$x^2 - c^2t^2 = x'^2 - c^2t'^2 \qquad \text{eq. 29}$$

But because y, y', z, & z' = 0 the Pythagorean process is not applicable. Therefore we remove the square exponents from eq. 29.

$$x - ct = x' - ct' \qquad \text{eq. 30}$$

Thus, we arrive back at our original equations for the **Figure 8** transformation. The only equation we can derive from eq. 30 is eq. 18 - which is only half of our eq. 2 derivation. We simply cannot derive eq. 1 from eq. 30.

Lorentz-Einstein both use the Galilean x-axis transformation equation and its respective inverse - adding γ to the rhs in both eq. 31 & eq. 32. But notice that not priming v for the term vt' in eq. 32 leaves the t in the denominator of v unprimed when we need to have only t' on the rhs and t on the lhs of eq. 32. See eq. 16 explanation above.

$$x' = (x - vt) \rightarrow x' = \gamma(x - vt) \qquad \text{eq. 31}$$

$$x = (x' + vt') \rightarrow x = \gamma(x' + vt') \qquad \text{eq. 32}$$

Einstein said c is constant in all reference frames so he substitutes $x = ct$ and $x' = ct'$ into eq. 31 and eq. 32.

$$x' = (x - vt) \rightarrow ct' = \gamma(ct - vt) \rightarrow ct' = \gamma t(c - v) \qquad \text{eq. 31}$$

$$x = (x' + vt') \rightarrow ct = \gamma(ct' + vt') \rightarrow ct = \gamma t'(c + v) \qquad \text{eq. 32}$$

He then multiplies (eq. 31) x (eq. 32), solves for γ and gets eq. 1. Lorentz stays with $x', x, vt, \& vt'$ and inserts eq. 31 and eq. 32 into squared eq. 26, 27, 28, & 29. And while his derivation is complex he arrives at eq. 1.

$$\gamma = \frac{1}{\sqrt{1-\frac{v^2}{c^2}}} \qquad \text{eq. 1}$$

Let's do Einstein's substitution of ct' and ct into eq. 31 & 32. Because the distance of the lhs terms equals the distance of the rhs terms in eq. 31 & 32 the time factor difference between t and t' can be determined without using γ. But again, is vt' in eq. 32 correct? Let's investigate.

$$x' = x - vt \rightarrow ct' = ct - vt \qquad \text{eq. 31}$$

$$x = x' + vt' \rightarrow ct = ct' + vt' \qquad \text{eq. 32}$$

Einstein said because S and S' are inertial either frame can claim they're stationary and it's the other one moving. Thus, he reasons, both S and S' must observe the same velocity v for the other frame. Hence the reason he (and before him, Lorentz) don't prime v for the S' frame. But because this reasoning fails to acknowledge the earth and its respective earth-stationary emf as a reference frame it's simply incorrect to say both S and S' can claim they're stationary and it's the other one moving. Both frames can claim they are inertial (not accelerating) but only S Red can claim he is emf-stationary (see above **emf-stationary - the fastest decay rate speed)**. Thus, because S' Green is moving relative to the earth-stationary emf; he will always have a slower time-rate relative to emf-stationary - ie - Red's time-rate. Thus, Green's velocity v' coefficient will be different than Red's velocity v coefficient because of their different time-rates. But this does not mean Green and Red are observing different separation velocities. Just like $1.0\ kmph$ and $0.621\ mph$ have different coefficients because of their different length units but are still the same velocity.

Side Note - Green's time-rate is a function of his velocity relative to the earth-stationary emf regardless if Red is present or not. Red and Green, like all particles of matter, are just tangible coordinate markers in the emf. Red being stationary relative to the emf just simply allows Green to know the time-rate of emf-stationary for relative comparisons.

So, because we are expressing eq. 32 from the S Red frame we need to use vt in eq. 32 and not vt'. See **Figure 8** transformation equations 14, 15, & 17 to confirm. So,

$$x = x' + vt \rightarrow ct = ct' + vt \rightarrow ct' = ct - vt \qquad \text{eq. 32}$$

Notice that eq. 32 equals eq. 31.

Thus, after removing the Pythagorean square exponents and unnecessary γ factor and using the correct vt term, I see no plausible way to derive a transverse kinetic time dilation eq. 1 from the longitudinal **Figure 8** transformation equations. To derive eq. 1 we need to use the **Figure 5** transformation and its respective eq. 11. To derive eq. 2 we need to use the **Figure 8 & 10** transformations and their respective 18 & 24 equations. Notice that equations 11, 18, & 24 are derived from the individual linear equations in their respective transformation.

All Transformation Equations - Our p-c length L when stationary can be expressed with ct. Thus, L on the lhs of eq. 9 can be replaced with ct which allows us to remove t_l.

$$t_{l}\frac{2L}{c} = \frac{L}{c+v} + \frac{L}{c-v} \rightarrow \frac{2ct}{c} = \frac{ct'}{c+v} + \frac{ct'}{c-v} \rightarrow t = \frac{t'}{1-\frac{v^2}{c^2}} \qquad \text{eq. 2}$$

Our task now is to show how the product of $v't' = vt$ when the coefficients of $v' \neq v$ and $t' \neq t$.

Reciprocal Corollary of the Lorentz Invariance Principle - When does $1 = 39.37$? When a $1m$ long board is measured in inches: $1m = 39.37 inches$. Note that changing units, coefficients, or frames does not change the physical length of the board. The physical properties of a particle or event are innate and therefore independent of the units, coefficients or frame chosen to quantify them. This is a reciprocal corollary of the Lorentz invariance principle which says the laws of physics must be the same in all frames for all observers. Thus, when Red and Green observe the same event, their quantification of the event's innate properties must be the same independent of their respective units, coefficients, or frame used to quantify them.

Thus, when Red observes Green moving relative to him at $0.5c$, Red naturally describes Green's dilated and contracted times in his respective S frame terms. This can make Green say Red moves faster than light (ftl). But this apparent-only ftl violation is necessary so that Red and Green calculate the same innate property quantities when observing the same event. Therefore, because Red's and Green's separation distance is an event, its innate distance property is independent of each frame's respective terms, units, and coefficients. Thus, the coefficients for Red's v & t terms can differ from Green's v' & t' terms as long as Red and Green calculate the same separation distance for each other so that $vt = v't'$.

Figure 8 Transformation - Thus, because eq. 19 says there is a time factor difference between Red and Green, and velocity is time dependent, Green in his respective S' frame can observe Red at a speed greater than c. Green can either leave his calculation alone or he can convert it back into Red's respective S frame coefficients by using eq. 33. Let's leave it alone and see if Green calculates the same separation distance as Red.

In **Figure 8 & 10** we see $v't' = vt$. Let's solve for v'.

$$v't' = vt \rightarrow v' = \frac{vt}{t'} \qquad \text{eq. 33}$$

Solve eq. 19 for t'

$$t = \frac{t'}{\left(1-\frac{v}{c}\right)} \rightarrow t' = t\left(1 - \frac{v}{c}\right) \qquad \text{eq. 34}$$

Replace t' in eq. 33 with the rhs of eq. 34.

$$v' = \frac{vt}{t'} \rightarrow v' = \frac{vt}{t\left(1-\frac{v}{c}\right)} \rightarrow v' = \frac{v}{\left(1-\frac{v}{c}\right)} \qquad \text{eq. 35}$$

If Red says Green's velocity is $0.9c$ then Green says Red's velocity is

$$v' = \frac{0.9c}{\left(1-\frac{0.9c}{c}\right)} = 9.0c' \qquad \text{eq. 36}$$

Priming c' tells us it's being expressed from Green's S' frame with his unit of time s' because $1s' \neq 1s$. Notice that Green's observed velocity of Red at $9.0c'$ is $10x$ faster than Red's observed velocity of Green at $0.9c$. This is because Green's $1s'$ is derived from 10 Kubicks while Red's $1s$ is derived from only 1 Kubick.

For $10s$ elapsed in Red's frame we get $1s'$ elapsed in Green's frame. Notice that Green's elapsed time of $1s'$ is $10x$ "slower" than Red's elapsed time of $10s$.

$$t' = 10s\left(1 - \frac{0.9c}{c}\right) = 1.0s' \qquad \text{eq. 37}$$

Green says Red's distance from him is

$$S = v't' \qquad \text{eq. 16}$$

So multiplying $9.0c'$ for v' and $1.0s'$ for t' we get the following distance.

$$S = v't' \rightarrow \left(9.0\frac{3x10^8 m}{s'}\right)(1.0s') = 9.0\left(3x10^8 m\right) \qquad \text{eq. 38}$$

Red says Green's distance from him is

$$S' = vt \qquad \text{eq. 21}$$

So multiplying $0.9c$ for v and $10s$ for t we get the following distance.

$$S' = vt \rightarrow \left(0.9\frac{3x10^8 m}{s}\right)(10s) = 9.0\left(3x10^8 m\right) \qquad \text{eq. 39}$$

Thus, because eq. 38 = eq. 39, Red and Green calculate the same separation distance while using their different but respective S and S' frame coefficients and thus obey the Lorentz invariance principle. See **Figure 11** to understand where the time factor difference of $10:1$ between Red & Green physically comes from. Green p-c moving at $0.9c$ relative to the earth-stationary emf displaces 10 Kubicks of emf while emf-stationary Red p-c displaces only one. Remember, a Kubick is one cube (side length $1ct$) of earth-stationary emf. Thus, Green's time-rate is $10x$ "slower" than Red's because his photon has to travel $10x$ the emf distance of Red's to complete his respective $1s'$. Hence, Green's v' coefficient will be $10x$ greater than Red's v coefficient while Green's time coefficient t' will be $10x$ smaller than Red's time coefficient t. Hence, as the coefficient of v' goes up, the coefficient of t' goes down in equal proportion thereby keeping $v't' = vt$. Thus, Red's $9.0c'$ observed by Green is not actually exceeding c but only appears to because Green has his respective time-rate out of proportion relative to the emf-stationary time-rate (ie Red) by a factor of $10:1$ (10 Kubicks : 1 Kubick) which naturally shows up in Green's frame respective coefficients.

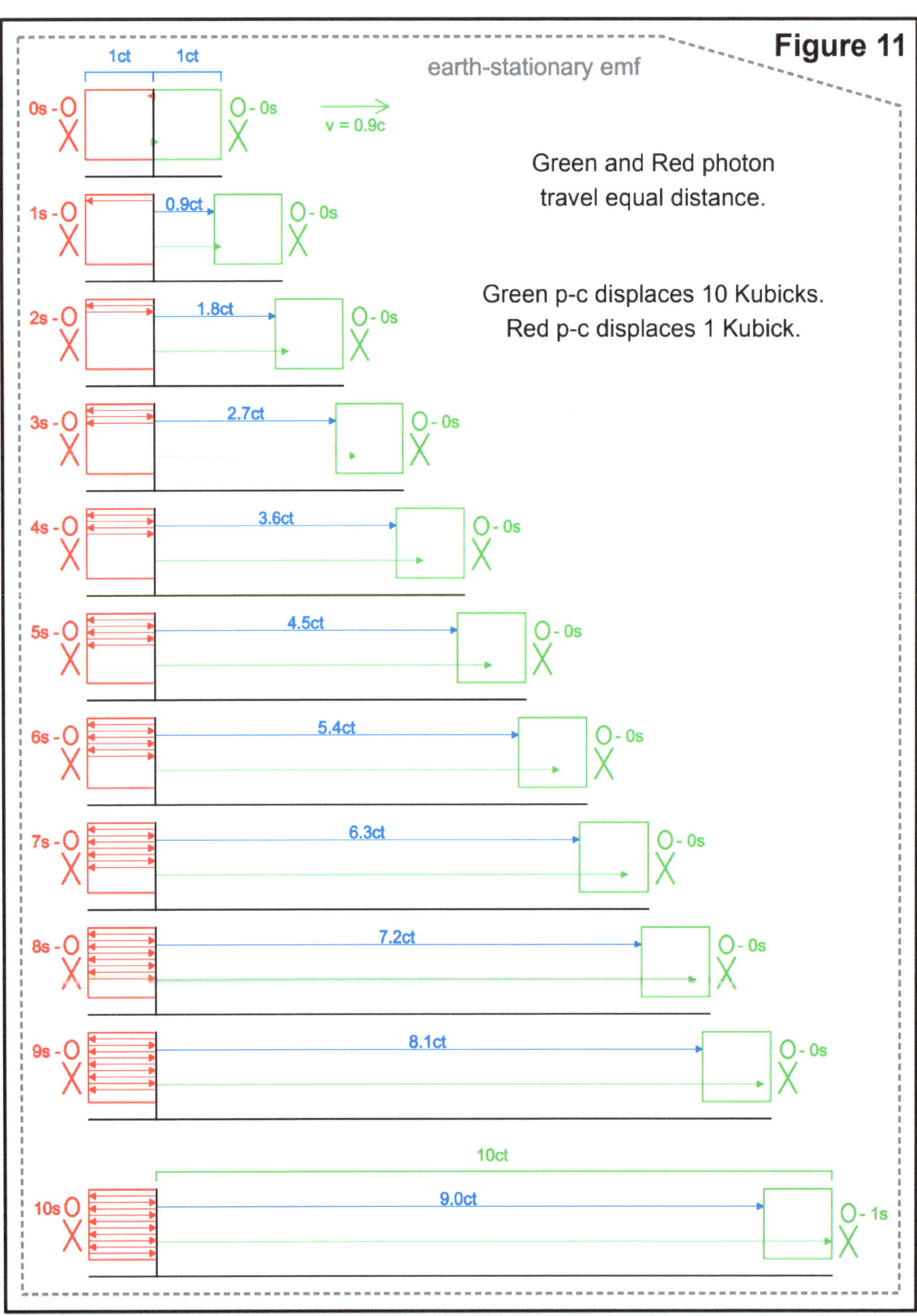

Figure 11
earth-stationary emf
1ct
1ct
0s - O
O - 0s
v = 0.9c
Green and Red photon
travel equal distance.
1s - O
0.9ct
O - 0s
2s - O
1.8ct
O - 0s
Green p-c displaces 10 Kubicks.
Red p-c displaces 1 Kubick.
3s - O
2.7ct
O - 0s
4s - O
3.6ct
O - 0s
5s - O
4.5ct
O - 0s
6s - O
5.4ct
O - 0s
7s - O
6.3ct
O - 0s
8s - O
7.2ct
O - 0s
9s - O
8.1ct
O - 0s
10ct
10s O
9.0ct
O - 1s

Let's replace vt with vt' in Green's separation distance eq. 38 to get eq. 40 and compare it to Red's separation distance eq. 39.

$$S' = vt' \rightarrow \left(0.9\frac{3x10^8 m}{s}\right)(1s') = 0.9\left(3x10^8 m\right) \qquad \text{eq. 40}$$

$$S' = vt \rightarrow \left(0.9\frac{3x10^8 m}{s}\right)(10s) = 9.0\left(3x10^8 m\right) \qquad \text{eq. 39}$$

For starters, eq. 40 is not possible because $s \neq s'$ and thus we can't divide them out for the same reason we can't take the product of tt'. And because eq. 40 ≠ eq. 39, Green calculates the wrong distance to Red and thus violates the Lorentz invariance principle whereby two different frames calculate different distances and coordinates for the same event.

Let's show how our unprimed and primed units divide out correctly to show velocities with different frame coefficients can be equal $9.0c' = 0.9c$. From our calculations above $v' = 9.0c'$, $t' = 1s'$, $v = 0.9c$, $t = 10s$.

$$\text{Green says } v' = \frac{vt}{t'} \rightarrow \frac{\left(0.9\frac{3x10^8 m}{s}\right)(10s)}{1s'} = \frac{9.0\left(3x10^8 m\right)}{1s'} = 9.0c' \qquad \text{eq. 41}$$

$$\text{Red says } v = \frac{v't'}{t} \rightarrow \frac{\left(9.0\frac{3x10^8 m}{s'}\right)(1s')}{10s} = \frac{0.9\left(3x10^8 m\right)}{1s} = 0.9c \qquad \text{eq. 42}$$

Notice we never prime m. Thus, we must conclude like c, m is invariant. If we are uncomfortable of letting Green speak his respective frame's coefficient language when he observes Red's velocity at $9.0c'$ then we can convert it back into Red's respective frame coefficient language using eq. 33.

$$v't' = vt \rightarrow v = \frac{v't'}{t} \rightarrow \left(\frac{\left(9.0\frac{3x10^{8}m}{s'}\right)(1s')}{10s}\right) = 0.9c \quad \text{eq. 44}$$

GPS satellite clocks undergo a similar time factor difference conversation so that their time coefficients match the time coefficients of clocks on earth.

Figure 10 Transformation - Let's confirm our **Figure 10** transformation.

In **Figure 10** we see $v't' = vt$. Solve for v'.

$$v't' = vt \rightarrow v' = \frac{vt}{t'} \quad \text{eq. 33}$$

Solve eq. 25 for t'

$$t = \frac{t'}{\left(1+\frac{v}{c}\right)} \rightarrow t' = t\left(1 + \frac{v}{c}\right) \quad \text{eq. 45}$$

Replace t' in eq. 33 with the rhs from eq. 45 above.

$$v' = \frac{vt}{t'} \rightarrow v' = \frac{vt}{t\left(1+\frac{v}{c}\right)} \rightarrow v' = \frac{v}{\left(1+\frac{v}{c}\right)} \quad \text{eq. 46}$$

If Red says Green's velocity is $0.9c$ then Green says Red's velocity is

$$v' = \frac{0.9c}{\left(1+\frac{0.9c}{c}\right)} = 0.4736842105c' \quad \text{eq. 47}$$

Notice that Green's observed velocity of Red at $0.4736842105c'$ is $1.9x$ slower than Red's observed velocity of Green at $0.9c$.

For $10s$ elapsed in Red's frame we get $19s'$ elapsed in Green's frame. Notice that Green's elapsed time of $19s'$ is $1.9x$ "faster" than Red's elapsed time of $10s$.

$$t' = t\left(1 + \frac{v}{c}\right) \rightarrow 10s\left(1 + \frac{0.9c}{c}\right) = 19s' \qquad \text{eq. 48}$$

Thus, Green says Red's distance from him is

$$S = v't' \qquad \text{eq. 16}$$

So multiplying $0.4736842105c'$ for v' and $19s'$ for t' we get the following distance.

$$S = v't' \rightarrow \left(0.4736842105\frac{3x10^8m}{s'}\right)(19s') = 9.0\left(3x10^8m\right) \qquad \text{eq. 49}$$

Red says Green's distance from him is

$$S' = vt \qquad \text{eq. 21}$$

So multiplying $0.9c$ for v and $10s$ for t we get the following distance.

$$S' = vt \rightarrow \left(0.9\frac{3x10^8m}{s}\right)(10s) = 9.0\left(3x10^8m\right) \qquad \text{eq. 50}$$

Thus, because eq. 49 = eq. 50, Red and Green calculate the same separation distance while using their different but respective S and S' frame coefficients thereby obeying the Lorentz invariance principle.

Eq. 2 - Let's confirm eq. 2 because it's a sum of **Figure 8** time dilation and **Figure 10** time contraction.

In **Figure 8 & 10** we see $v't' = vt$. Solve for v'.

$$v't' = vt \rightarrow v' = \frac{vt}{t'} \qquad \text{eq. 33}$$

Solve eq. 2 for t'

$$t = \frac{t'}{\left(1 - \frac{v^2}{c^2}\right)} \rightarrow t' = t\left(1 - \frac{v^2}{c^2}\right)$$ eq. 2

Replace t' in eq. 33 with the rhs from eq. 2 above.

$$v' = \frac{vt}{t'} \rightarrow v' = \frac{vt}{t\left(1 - \frac{v^2}{c^2}\right)} \rightarrow v' = \frac{v}{\left(1 - \frac{v^2}{c^2}\right)}$$ eq. 51

If Red says Green's velocity is $0.9c$ then Green says Red's velocity is

$$v' = \frac{0.9c}{\left(1 - \frac{0.9c^2}{c^2}\right)} = 4.736842105c'$$ eq. 52

Because $v \,\&\, c$ are squared in eq. 2 due to it being derived from the sum of **Figure 8** time dilation and **Figure 10** time contraction, we will no longer see a linear difference between $S \,\&\, S'$ frame coefficients. At this point it's easier to call their coefficient differences their factor difference.

Green's observed velocity of Red at $4.736842105c'$ has a factor difference of 0.19 relative to Red's observed velocity of Green at $0.9c$.

For $10s$ elapsed in Red's frame we get $1.9s'$ elapsed in Green's frame. The time factor difference between Green and Red is 0.19.

$$t' = t\left(1 - \frac{0.9c^2}{c^2}\right) \rightarrow 10s\left(1 - \frac{0.9c^2}{c^2}\right) = 1.9s'$$ eq. 53

Thus, Green says Red's distance from him is

$$S = v't'$$ eq. 16

So multiplying $4.736842105c$ for v' and $1.9s'$ for t' we get the following distance.

$$S = v't' \rightarrow \left(4.736842105 \frac{3x10^8 m}{s'}\right)(1.9s') = 9.0\left(3x10^8 m\right) \qquad \text{eq. 54}$$

Red says Green's distance from him is

$$S' = vt \qquad \text{eq. 21}$$

So multiplying $0.9c$ for v and $10s$ for t we get the following distance.

$$S' = vt \rightarrow \left(0.9 \frac{3x10^8 m}{s}\right)(10s) = 9.0\left(3x10^8 m\right) \qquad \text{eq. 55}$$

Thus, because eq. 54 = eq. 55, we see eq. 2 correctly expresses the sum of **Figure 8** transformation time dilation and **Figure 10** transformation time contraction as a longitudinal moving p-c at velocity v relative to the earth-emf-stationary frame.

Furthermore, because we've shown that time is directly proportional to the cycle displacement of c in a moving transverse and/or longitudinal p-c, we see time is a principle of displacement ratios. See **Figure 3** where Green's $1s'$ is $2ct$ long while Red's $1s$ is $1ct$ long. See **Figure 11** where Green's $1s'$ is $10ct$ long while Red's $1s$ is $1ct$ long.

Principle of Displacement Ratios - Our measurement of time is actually just a principle of displacement ratios where we simply measure the movement of something relative to a chosen something else moving at its own constant speed. That something else being the earth's rotation and orbit. Though we call it a clock, a sundial is actually just a degree wheel measuring how far the earth rotates relative to the sun. Our unit of time, $1s$, is literally a measure of how far the earth rotates in degrees relative to the sun. We take one earth rotation and divide it into our arbitrarily chosen units of "time". We then multiply these units by our arbitrarily chosen quantity of degrees for a circle.

$$1s = 1rev/24hrs\ x\ 60min\ x\ 60s = 86,400\ x\ 360\ degrees = 0.00416667deg$$

So when the elapsed time of an event is $1000s$ what we're really saying is that the net displacement of the event is equal to the earth rotating $4.16667\ degrees$ $(1000\ x\ 0.00416667degrees)$. Thus, our notion of time reduces to a principle of displacement ratios - ie - how much displacement did an event undergo relative to the rotational displacement of earth. Why is everything in the universe always in constant displacement? Who knows, but it prevents the universe from running out of time.

Spatial Rate - In the denominator of eq. 4 it appears we're subtracting v from c and in the denominator of eq. 6 it appears we are adding v to c and thus doing velocity subtraction and addition with c. But we're not. Because c and v are two separate reference frames moving independent of and relative to each other it's not possible for them to undergo velocity subtraction or addition. To understand what is actually happening we need to explain relative velocity with spatial quantity. See the explanations in **Figures 12 & 13**.

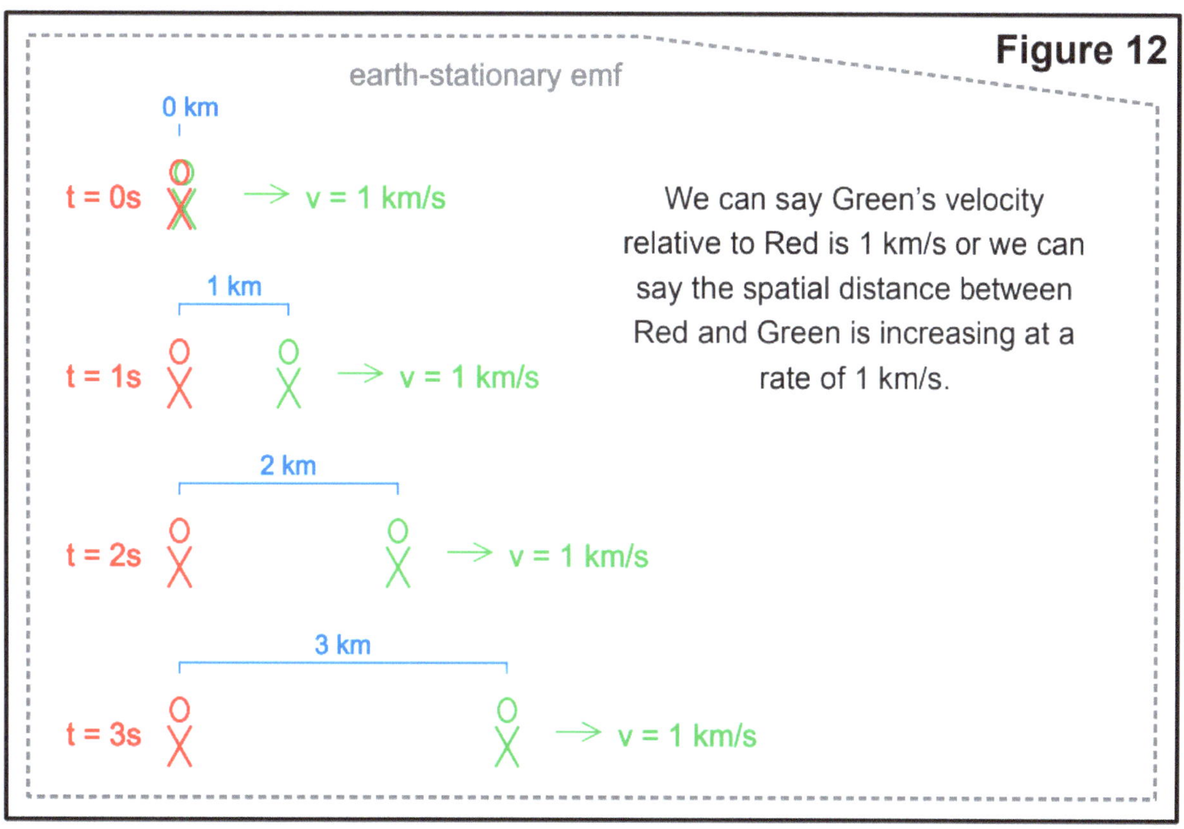
Figure 12
earth-stationary emf
0 km
t = 0s
v = 1 km/s
We can say Green's velocity relative to Red is 1 km/s or we can say the spatial distance between Red and Green is increasing at a rate of 1 km/s.
1 km
t = 1s
v = 1 km/s
2 km
t = 2s
v = 1 km/s
3 km
t = 3s
v = 1 km/s

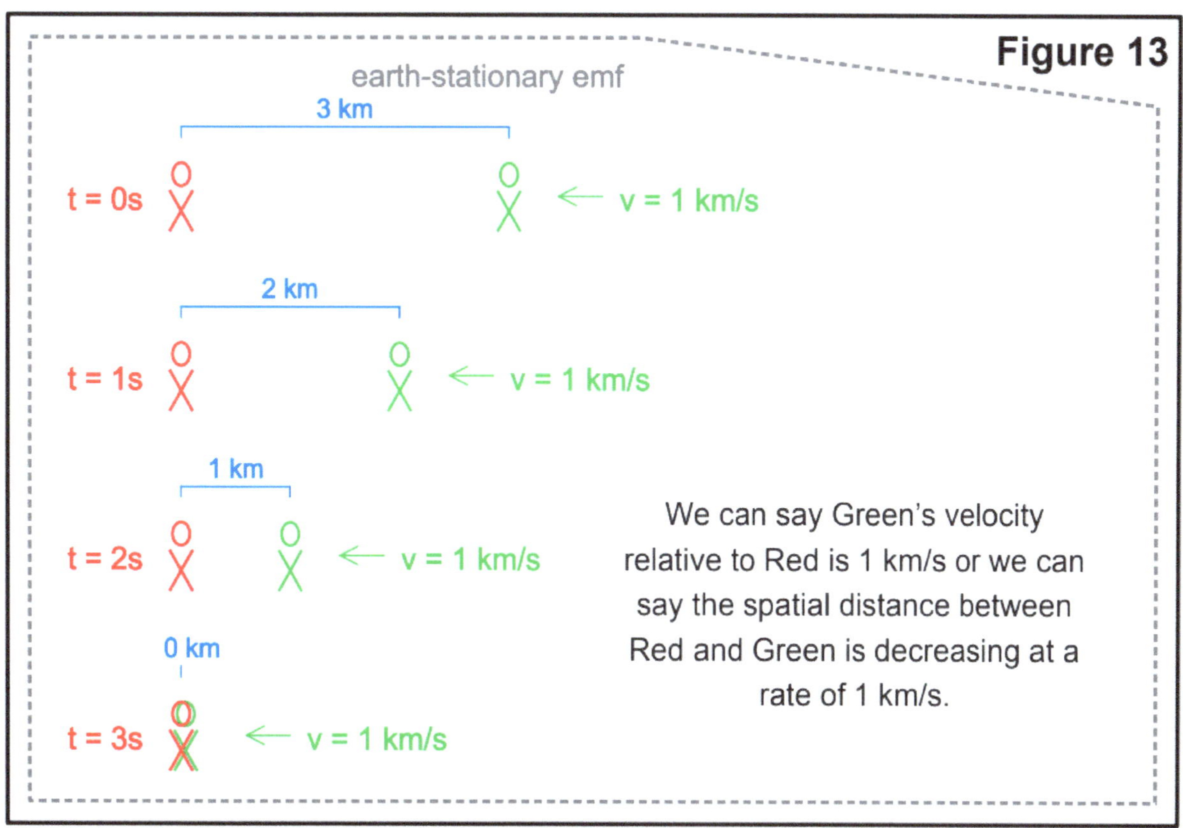
Figure 13
earth-stationary emf
3 km
t = 0s
v = 1 km/s
2 km
t = 1s
v = 1 km/s
1 km
t = 2s
v = 1 km/s
We can say Green's velocity relative to Red is 1 km/s or we can say the spatial distance between Red and Green is decreasing at a rate of 1 km/s.
0 km
t = 3s
v = 1 km/s

Thus, the denominator of eq. 4 says c is advancing towards its eventual i-vent with the p-c's right wall at velocity c while the wall simultaneously moves away from c at $0.5c$ (see **Figure 3**). Thus, the spatial distance between c and the right p-c wall is decreasing at a rate of $0.5c$ $(c - v)$.

The denominator of eq. 6 says c is advancing towards its eventual i-vent with the p-c's left wall at velocity c while the wall is simultaneously moving towards c at $0.5c$ (see **Figure 4** after the right i-vent). Thus, the spatial distance between c and the oncoming left p-c wall is decreasing at a rate of $1.5c$ $(c + v)$.

Boat & Lake Analogy - Because matter can move relative to the (emf) and its photons (Doppler effect) the path of a photon is independent of its source. In other words, a particle's vector does not carry-over to the vector of the photon it emits. This is because the photon leaves the mass/Higgs field (matter) and emerges in the emf when emitted and thus goes where the emf field dictates. To understand this, imagine you're in a boat sitting stationary relative to the lake, and your friend, in his boat, passes you at 100 mph (the boats are analogous to the mass/Higgs field). As your friend passes you, you and him each drop a rock in the water at the same time at slightly different points (the water is analogous to the emf). Ignoring amplitude, the resulting waves from each dropped rock will propagate outward with the same speed and circular expansion. This is because the waves are a property of the water and independent of the vector of either boat or rock. This same principle applies to a photon-emitting particle; the speed and path of an emitted photon (minus the biased act of aiming) is a property of the emf and independent of its source's vector. In other words, light cannot travel sideways or slant-ways because the electric and magnetic wave components of a light wave must oscillate transversely to the light-wave's line of propagation. Thus, a photon emitted vertically up the y-axis at $(0, 0)$ from a particle moving along the x-axis at velocity v will not deviate from $x = 0$ as it propagates up the y-axis. **Light-time correction** corroborates this phenomenon.

A & B - Imagine me on planet A separated from you on planet B so that it takes c $1s$ to go from the surface of A to the surface of B as measured by me. Both A and B are stationary relative to each other and the emf. I observe the elapsed time of c for the A-B-A cycle distance at $2s$. You on the smaller less massive B have a faster time-rate because of gravitational time dilation observe the elapsed time of c for the same inverted B-A-B cycle distance at $4s$. Because $2s \neq 4s$ do we conclude the distance B-A-B is greater than A-B-A? No. Do we conclude that c is slower on B than A? No.

$$A - B - A = D \rightarrow c_A = \frac{2D}{2s} = 1c_A \qquad \text{eq. r1}$$

$$B - A - B = D \rightarrow c_B = \frac{2D}{4s} = 0.5c_B \qquad \text{eq. r2}$$

Because we know c is constant and of gravitational time dilation we see B's time runs faster than A's by a factor of $2x$. Thus, if A is our frame of reference to measure all other frames against, then we multiply eq. r2 by our factor difference of $2x$ so $c_A = c_B$. This will allow B to make correct calculations for events on A just like GPS satellite clocks are time factor corrected in order to correctly calculate event locations on earth. But note that B could never be aware of A's time-rate and still correctly predict observed events on A as long as B keeps everything in his respective B frame units. We do it all the time from earth for observed events in our universe.

Postamble - After the 1887 Michelson-Morley experiment failed to detect the aether there was some debate to whether or not to conclude the aether was stationary relative to earth. This would have explained the lack of interference or null result of the experiment. But it was Einstein with Special Relativity in 1905 who assumed there was no aether and that all inertial frames, independent of velocity, can claim they are stationary. Here is the reciprocal corollary of this assumption. All frames moving relative to one another can claim they are the one moving and it is the other one that is stationary.

Is earth and its respective emf a frame? Can they claim they're at rest? Since we have replaced the aether with the emf we see evidence that the emf surrounding earth is stationary relative to the earth while both orbit the sun at $30\ kmps$. The 1887 M-M null result experiment says earth's emf is earth-stationary as do the GPS satellite clocks moving relative to earth at $\sim 3.9\ kmps$ undergoing kinetic time dilation. Thus, if we were to take a Michelson interfermoeter and give it a velocity v relative to earth, would it produce an interference pattern?

thx - God Bless.

www.ingramcontent.com/pod-product-compliance
Lightning Source LLC
LaVergne TN
LVHW071130160826
845679LV00005B/1230

* 9 7 9 8 8 2 1 0 8 3 9 9 9 *